Brilliant Valentine Science Experiment Ideas

Valentine's Day Science Activities to Wow The Kids

Copyright © 2021

All rights reserved.

DEDICATION

The author and publisher have provided this e-book to you for your personal use only. You may not make this e-book publicly available in any way. Copyright infringement is against the law. If you believe the copy of this e-book you are reading infringes on the author's copyright, please notify the publisher at: https://us.macmillan.com/piracy

Brilliant Valentine Science Experiment Ideas

Contents

Valentine Slime 1

Heart Magic Milk Experiment 7

Borax Crystal Hearts 13

Valentine's Day Lava Lamp Science Experiment 19

Valentines Oil And Water Science Density Activity 26

Valentine's Day Volcano Science Experiment 33

Valentine Bubble Science 44

Valentine Slime

If you have tried some of our other slime recipes then you know we love to make slime here! And slime can be a great non-candy gift idea for Valentine's Day. Kids can have fun helping you to make the slime and printing off these cute slime gift tags "Happy Valen-slime Day". This Valentine Slime will make a great gift this Valentine's Day!

Ingredients Needed to Make Valentine Slime

– 1 Bottle of Elmer's Glitter Glue (6 oz). We used red and pink. Note: We have only tested this with Elmer's glitter glue and recommend this brand as other brands may not work.

– 1 TBSP or more (up to 1 cup) water which will make your slime stretchier.

– Red & Pink Glitter

– 1/2 TBSP Baking Soda

– 1½ TBSP of Contact lens solution. **Important: your brand of contact lens solution must contain boric acid and sodium borate. See our full contact solution slime recipe for the brands we recommend.

To Make the Slime Jars You'll Need:

–Small Mason Jars

–Cardstock for printing the gift tags (grab the gift tags at the bottom of this post)

–Red & White Twine

Brilliant Valentine Science Experiment Ideas

Directions to Make Valentine Slime

We made two glitter glue slimes – this was the red slime.

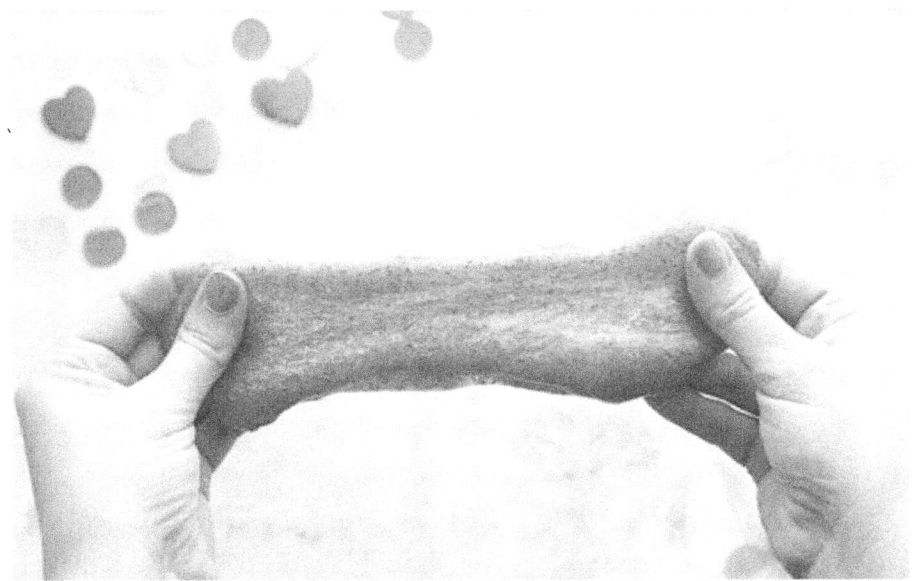

And this is the pink slime.

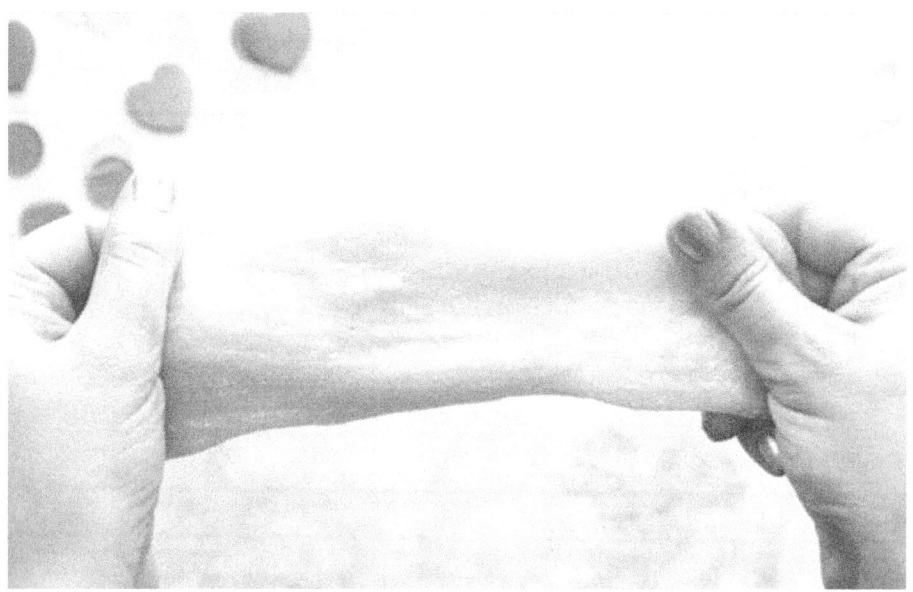

Once you have both slimes made, you can fill your small mason jars. Depending on your size of jars you should be able to get 2-3 jars with one batch/recipe. We filled 2 small mason jars with one batch.

Now add your gift tags with a little bit of twine wrapped around the jar!

Adult supervision is recommended and slime should never be placed in the mouth. Adults should make the slime. See the full recipe for more slime safety tips.

Heart Magic Milk Experiment

The first experiment we're going to show you is how to do a heart magic milk experiment.

Here is what you will need:

Milk – high fat milk works better (we used 3.25%)

Dish soap

Food Coloring – red and pink

Heart-shaped cake pan or regular pan

Cotton swabs or cotton balls

Steps to do a Magic Milk Experiment

1. First pour some milk into your heart-shaped cake pan. You don't need that much milk, just a small layer.

2. Add some drops of food coloring into your milk.

3. Now dip a cotton swab into some dish soap. You can also soak a cotton ball. Gently tap the cotton swap onto your food coloring or drop your cotton ball in. Now watch the magic!

The science: Milk has tiny amounts of fat. The secret to the explosion is the soap. When the soap is added to the milk, we are seeing the soap molecules moving around trying to join up with the fat molecules. The food coloring molecules get moved around in the process, showing us what is happening between the milk fat and the soap. Once the soap and milk gets evenly mixed the movement of the food coloring stops. Higher fat milk will produce more explosion because there is more fat to mix with the soap.

Heart Skittles Experiment

Brilliant Valentine Science Experiment Ideas

Our second Valentine's Day science experiment is a skittles candy heart. In this experiment, you can observe candy dyes dissolving but also that the dyes won't mix with each other as they dissolve.

Here is what you will need:

Package of skittles

Warm water

Plate that is slightly curved in

Steps to do a Skittles Experiment

1. Line up your skittles into the shape of a heart on a dish that is slightly curved in. It needs to be curved so that the candy dye will run

into the center. You can use red and pink or make a rainbow of colors. We loved the way the rainbow experiment turned out!

2. Slowly add in warm water into the center of your heart. Try not to get the water beyond the skittles.

Now watch the magic! The candy dye will slowly dissolve and the dye color will move from the candy to the center of the dish.

The science:

Here's what you'll observe: the food dye from the skittles doesn't initially mix with the water. The skittles dissolve, but the colors do not initially mix with each other. The reason the food colours meet in the middle of the plate and do not initially mix is because each skittle has the same amount of sugar dissolving.

Follow-up Experiments: Ask kids what they think will happen and what their predictions are when you change the experiment. Try changing variables like adding cold water, different candy or adding in sugar cube obstacles to see what happens.

Brilliant Valentine Science Experiment Ideas

Borax Crystal Hearts

Our final Valentine's day science experiment is growing borax crystal hearts. These will start to form within 24 hours and it's a really fun science experiment for kids to observe!

Here is what you will need:

A mason jar

Borax (adults only to handle)

Boiling water

Red pipe cleaners and craft sticks

1. Add 3 tablespoons of borax powder for every 1 cup of boiled water into a mason jar. Stir to make sure all of the borax dissolves in your water.

Adults only should do this part – the boiled water in the jar will be very hot and borax is a chemical that only adults should handle. Please keep borax out of reach from young children.

2. Make a heart from a pipe cleaner. Use another pipe cleaner to twist it onto a popsicle stick. The popsicle stick will rest on top of your mason jar to allow the heart to hang inside the solution. Make sure your heart is small enough to fit into the mason jar with some extra space for where the crystals will grow. This is so that you can

remove the crystal heart from the mason jar once the crystals are fully formed.

3. Add your heart pipe cleaner into your jar with borax solution and let it sit for 24 hours. Check in throughout the day to see the progress of your crystals growing on the pipe cleaner!

4. After 24 hours you should have fully formed crystals on your heart! You may even get some crystals formed on the bottom of your jar.

Brilliant Valentine Science Experiment Ideas

The science: When you mix the borax powder with hot water, the hot water molecules are moving really fast which allows more borax to dissolve into it. When the water mixture starts to cool, these water molecules slow down and then move closer together. This leaves less room for the dissolved borax and it begins to separate out of the water. As the borax "falls" out of the water mixture it bonds with other borax and will start to crystallize on the pipe cleaner heart. The borax continues to crystallize on the heart until you remove it out of the water.

Follow-up experiments: See what happens when you run the experiment again with your crystal heart. Do more crystals form on top? Or if you leave the heart in for 48 hours, will you get a larger crystal heart? Will the crystals form if you try to dissolve the borax with lukewarm or cold water?

Working with Borax Safety Reminders:
- Adults only should handle the borax. Always read and follow the labels of products used.
- Borax can be harmful if swallowed, inhaled or if it gets into your eyes.
- To minimize borax dust from spreading in the air when doing this experiment, always add the water first into the container and then add the borax.

We hope you enjoy these 3 really easy Valentine science experiments!

Valentine's Day Lava Lamp Science Experiment

Learn about science this Valentine's Day with this fun Valentine's lava lamp science experiment! Kids will have a blast with this Valentine's Day STEM activity!

You know what my kids love? Things that fizz and bubble. They are delighted to play with anything that is fizzy. One of our favorite activities are lava lamps, not only because they are fizzy, but also because they are an easy way to show kids how gasses can move through liquids and make bubbles.

The lava lamp science experiment is a fun way to explain how their burps work, and if your kids are anything like mine, they are oddly fascinated by burping.

What you'll need for the Valentine's Day lava lamp science experiment:

Science beaker (or you can use a cup or jar)

Pink food coloring

White glitter

Alka seltzer tablets

Water

Vegetable oil or baby oil

First, fill the beaker 3/4 of the way full with oil. Add about a 1/3 inch line of pink water to the cup.

Add white glitter to the jar for extra sparkle. You can also add heart confetti.

Brilliant Valentine Science Experiment Ideas

Place the jar on a surface where kids can look at it at eye-level.

Drop in a quarter of an Alka seltzer tablet into the jar. Watch the bubbles fly high!

The more tablet you put in, the more vigorously it will boil.

You can repeat this activity for as long as you have tablets!

Lava Lamp Science

Oil is less dense than water, so it remains on the surface of the colored water. Additionally, inter-molecular polarity prevents oil and water from mixing (oil molecules are not attracted to water molecules). But, when you add alka seltzer tablets, which contain citric acid and baking soda that react when placed in water, creating carbon dioxide. The gas is less dense than either the water or the oil, but the gas bubbles take some of the water with them. Then, when the gas rises to the top of the oil, the heavier water drops back down, creating a bubbling lava lamp effect.

Valentines Oil And Water Science Density Activity

There are many simple science experiments that can be done right on the kitchen counter. This Valentines Oil and Water Science Activity is one of them! We love holiday themed science ideas.

Turn basic science into Valentine's Day science with easy to set up themes. Think classic Valentine's colors like red, pink, and purple. Heart cookie cutters, heart gems, and even little heart containers make this activity extra special.

Our simple Valentines oil and water science activity are lots of fun! We have been exploring a bunch of quick Valentine's Day themed science activities and experiments this season. So far we have tried out Valentine's Day themed viscosity, baking soda eruptions, water displacement, and bubble science activities.

SUPPLIES

Baby Oil (regular cooking oil will work but it won't be clear)

Eyedropper Straws (for blowing the water if desired)

Red Colored Water

Heart Shape Cookie Cutters

Plastic heart gems (dollar or craft store)

Tray or Lid to Storage Bin (for experimenting and mess collection)

VALENTINES DAY OIL AND WATER SET UP

Grab a lid or tray.

Next, fill a cup with water and added red food coloring. Then add at least 15 drops to get it really red! Why not try all the colors.

Set out the tray, an eyedropper, and any kind of Valentine's Day themed items you want to use like cookie cutters or heart-shaped table scatter!

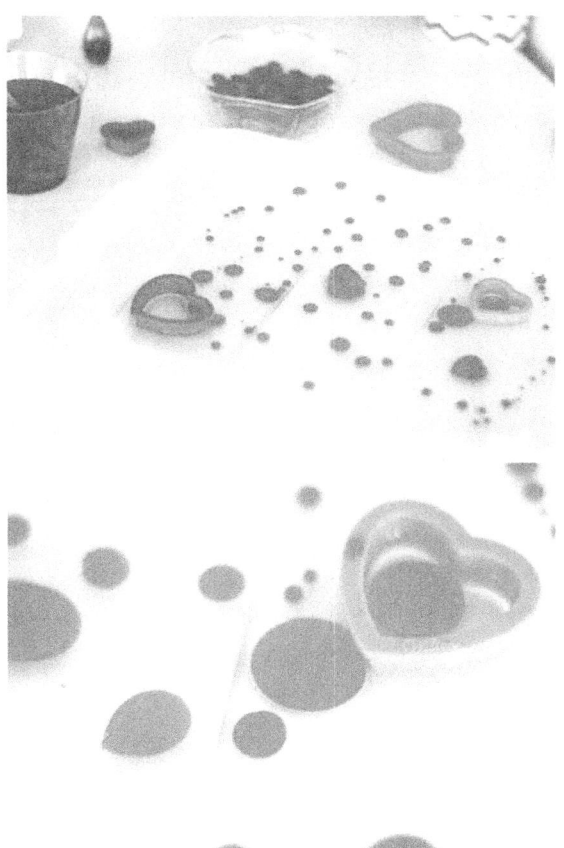

OIL AND WATER VALENTINES DAY SCIENCE INFO

This Valentine's oil and water science activity is all about density! The density of liquids. If you were to put oil and water into a jar, the two would not mix.

The two liquids separate because oil is less dense than water.

Similarly, when you mix the oil and water in this activity, they don't mix. You can do this Valentines oil and water activity opposite too. Drop the oil onto the water!

You might recognize this as being similar to an oil spill in the ocean. The oil will sit on top. Try making a science discovery bottle with oil and water.

Science can be very playful for kids! This Valentines oil and water experiment is a great way to explore science while having fun. There aren't complicated steps to make kids frustrated. Science at home can be super easy!

Definitely hands-on learning for young kids! Check out how you can fill up the heart-shaped cookie cutters.

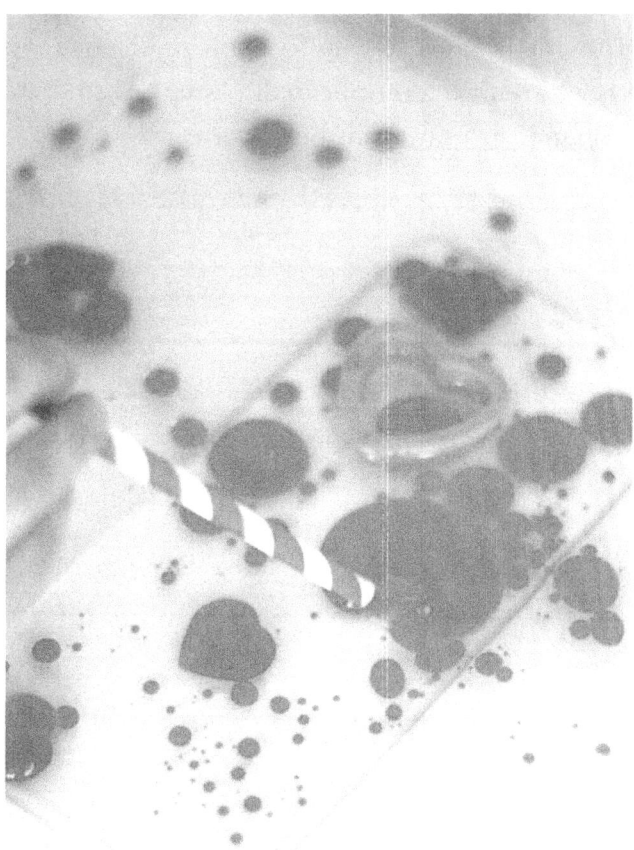

There's no wrong way to play with this *Valentines oil and science activity*. Playtime that involves exploring, experimenting, observing, and discovering is perfect for young kids. Early childhood science is the best for reinforcing a love for science.

Valentine's Day Volcano Science Experiment

It's that time of year again, the time to start thinking about Valentine activities, valentine cards, and trying to keep candy at a minimum. Try to come up with a couple of cool Non-Candy Valentine's Day ideas that the kids will love. This year it's a Valentine's Day Volcano Science Experiment. Every child loves seeing the reaction of baking soda and vinegar so why not pretty it up for a Valentines Day Science experiment.

What you will need to make your Valentine Volcano and Science Valentine's Day Cards is in this picture below.

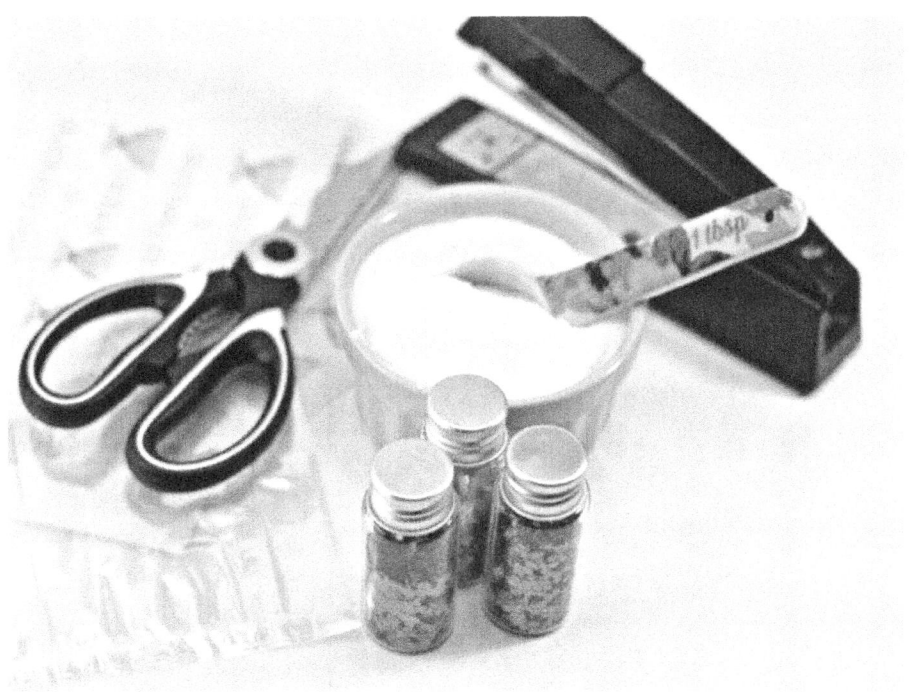

Kids love hands-on Science Activities, and this **Volcano Science Experiment** is one of our favorites. The great thing about this idea is you can use it as a fun non-candy Valentine gift.

Baking Soda and Vinegar Science Experiments are so easy and they'll excite your kids and keep them interested in science.

As you can see all you need is a few ingredients. Begin by placing the clear vase or container into a pan or dish to catch the mess that will come from the overflowing volcano. This will help to catch the mess when the fizzing effect overflows, and it makes for easy cleanup. WE ALL love easy cleanup.

Brilliant Valentine Science Experiment Ideas

To make these for a Valentine gift idea pour some baking soda into a bowl and add glitter or Valentine confetti.

Then separate the baking soda mixture into individual snack size bags or vials. You can see the picture below for an example of how we set ours up.

Download and Print out the free Valentine's Day Cards

Add your baking soda mixture into the bag, then staple your Valentine card on the bag.

The children will need to add the vinegar to create their eruption

Brilliant Valentine Science Experiment Ideas

For the vinegar, we like to add a drop of red or pink food coloring to keep with the Valentine theme.

Brilliant Valentine Science Experiment Ideas

Volcano Experiment for Kids

Valentines Day Science

Begin by placing the clear vase or container into a pan to catch the mess that will come from the overflowing volcano.

Brilliant Valentine Science Experiment Ideas

Add some baking soda and Valentine glitter or confetti into your clear container (vase)

Place a drop or two of pink food coloring into some distilled vinegar if you would like your liquid volcano to be colored.

Pour the vinegar into the container with the baking soda and watch it erupt in all of its pink glitter-filled glory

Pour the colored vinegar into the vase over top of the baking soda.

Watch as a fizzing reaction occurs and bubbles over. Adding color to your vinegar is optional but it does make for a nice Valentine explosion.

Are you and your kids ready for some Valentine's Day Science fun? This Fizzy Baking Soda Science idea will be the biggest hit amongst the kids'. Whether you do this for a classroom, your own children, or your child hands this Valentine card gift idea out to friends I'm sure it will be a success.

If you are looking for simple Science experiments and activities, add this one to your bucket list. Kids love bubbly, fizzy, erupting, colorful fun. Which makes this Valentine Volcano Activity perfect every time you do it.

Valentine Bubble Science

This fun Valentine bubble science activity for preschool is perfect for engaging young kids and getting them interested in science! Young children love things they can see and touch. Using bubbles at home or in the classroom is a great way to naturally provide opportunities for open-ended inquiry and exploration- important scientific skills that will set the stage for future success in school and life.

Science with holiday themes allows you to explore a prior science experiment or activity in a new way.

We used a fun Valentine's Day tray, straws, and cookie cutters from the dollar store for this bubble science activity.

You can see how it can easily be changed for seasons and holiday. Plus, it's a great science activity all year long!

Supplies Needed:

Water

Dawn Dish Soap {this brand works best}

Light Corn Syrup

Measuring Cups

Spoons

Themed Accessories

Use the recipe below to mix up a batch of bubble solution. You can make less if you need to. We made a ⅓ of the recipe just for us.

Brilliant Valentine Science Experiment Ideas

BUBBLE SCIENCE
RECIPE FOR BEST BUBBLES

3 Cups Water

1 Cup Dish Soap

1/2 Cup Light Corn Syrup

measure| pour| stir gently

If possible, have the kids help with the bubble solution mixture making!

Recipes are great math lessons for kids. Pouring ingredients also helps develop fine motor and practical life skills.

Bubble Science

Encourage kids to blow bubbles into the tray while you share some information about the science behind the bubbles! Here are a few suggestions:

Soap forms the thin wall or skin of the bubble

This thin layer traps air inside creating the bubble

When two bubbles meet they join, sharing one common wall

When two bubbles of the same size meet they become one

When the water evaporates, the bubble pops

There are so many more science facts about bubbles, but this is a good starting point.

Brilliant Valentine Science Experiment Ideas

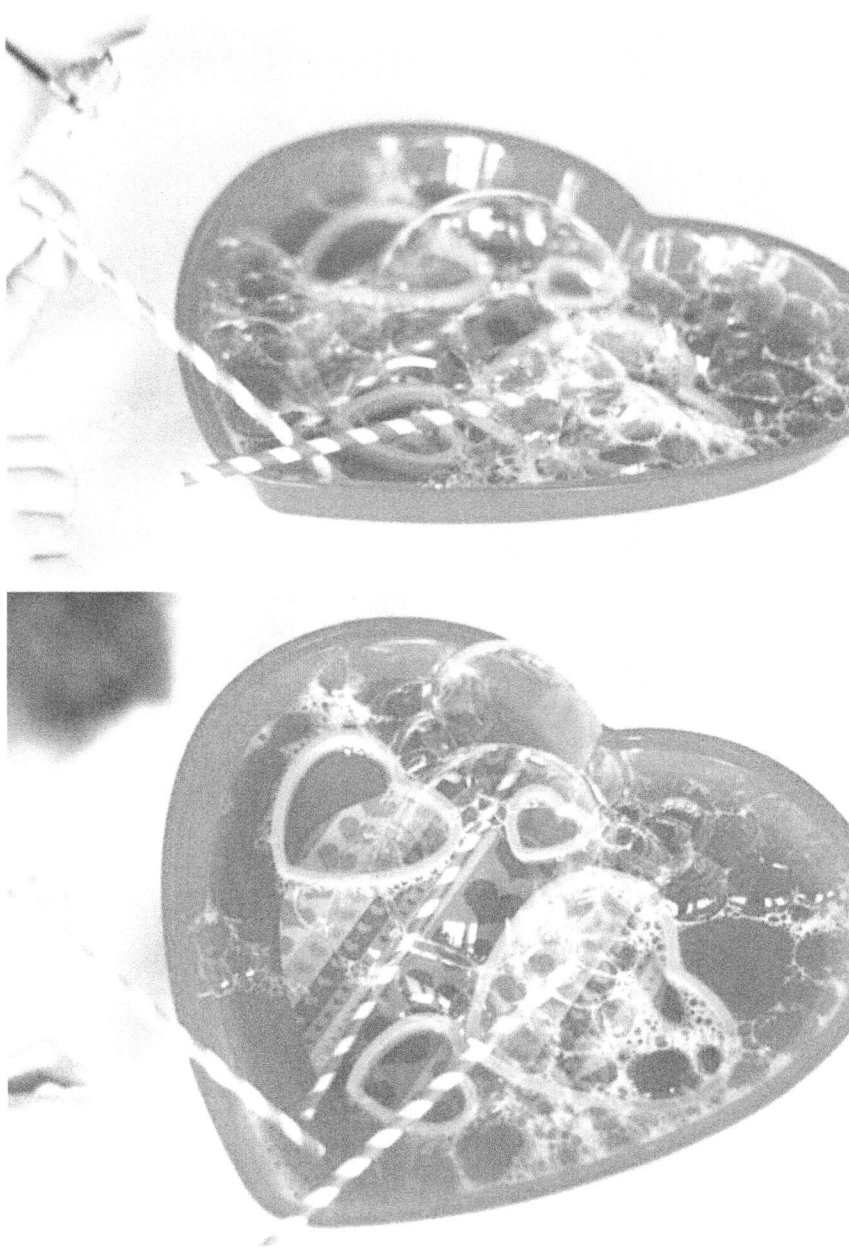

Making Bubbles

You can encourage experimentation and play with the bubble solution. Here are a few questions you can ask your pre-K or kindergartern students:

Can you use the cookie cutters to blow bubbles?

Can you use the end of the straw to blow bubbles?

Can you make a bubble tower {provide small cups of bubble solution}?

Can you use the straw to blow bubbles inside the cookie cutters?

Can you push your straw into the bubble without it popping? (Trick: dip the straw into the solution first to create a barrier.)

Can you hold a bubble without it popping? (Trick: Make your hands soapy first!)

Brilliant Valentine Science Experiment Ideas

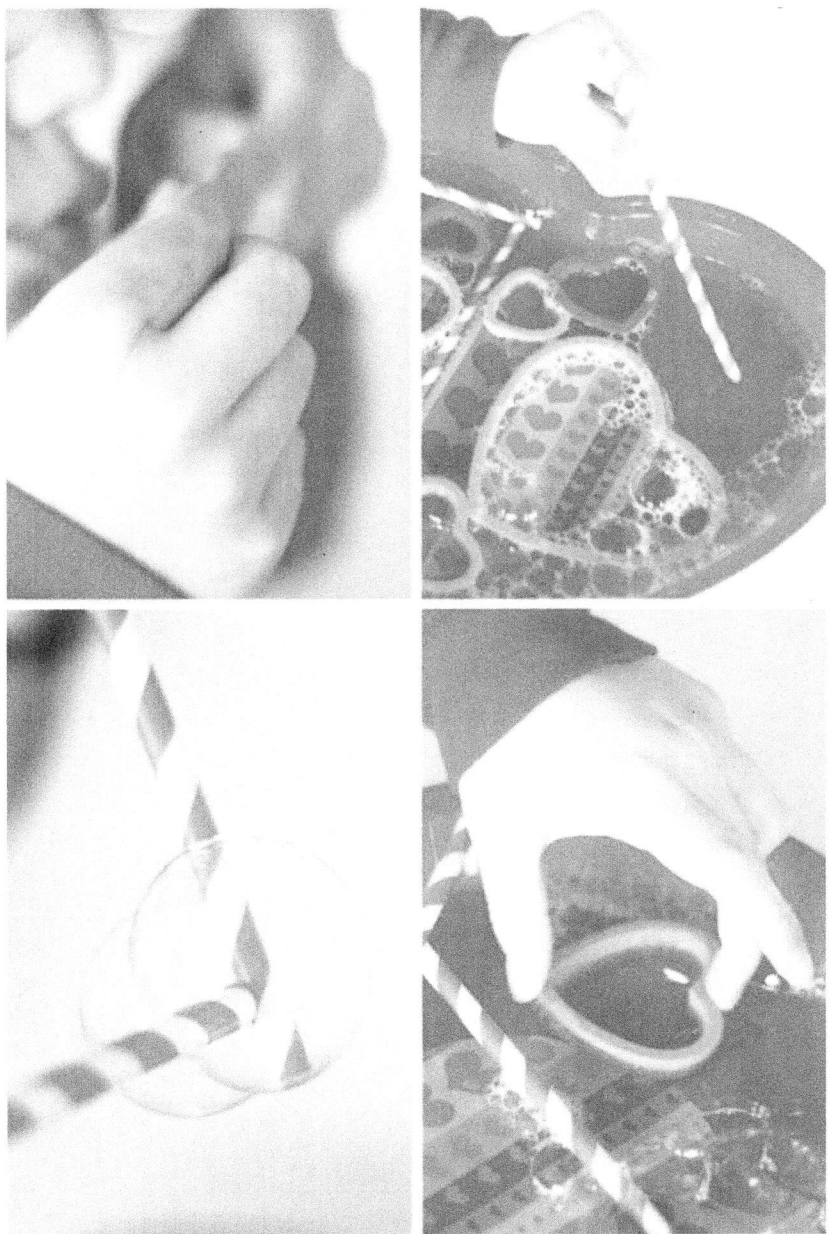

51

What happens when you blow hard and fast into the straw?

Or what happens when you blow slowly and softly into the bubble solution?

Are the bubbles different?

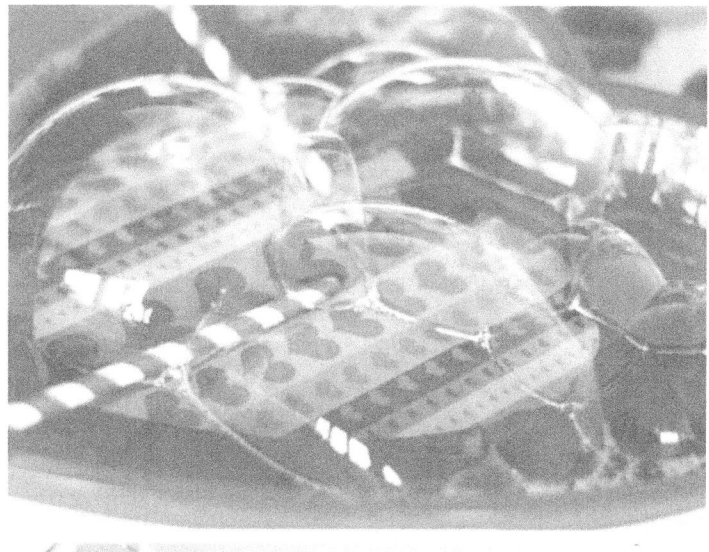

Blow bubble towers as a grand finale! Complete your Valentine bubble science activity with bubble towers! How big or tall can you make your tower? Provide each kid with a plastic cup and a straw!

Made in United States
Orlando, FL
10 February 2024

43528174R00039